YOUR KNOWLEDGE HAS VALUE

- We will publish your bachelor's and
  master's thesis, essays and papers

- Your own eBook and book -
  sold worldwide in all relevant shops

- Earn money with each sale

Upload your text at www.GRIN.com
and publish for free

Amalia Aventurin

Thermal conductivity scanner (TCS)

Petrophysics

GRIN Verlag

Bibliografische Information der Deutschen Nationalbibliothek:

Die Deutsche Bibliothek verzeichnet diese Publikation in der Deutschen National-
bibliografie; detaillierte bibliografische Daten sind im Internet über http://dnb.d-
nb.de/ abrufbar.

Imprint:

Copyright © 2013 GRIN Verlag GmbH
Druck und Bindung: Books on Demand GmbH, Norderstedt Germany
ISBN: 978-3-656-64481-1

This book at GRIN:

http://www.grin.com/en/e-book/272603/thermal-conductivity-scanner-tcs

Thermal conductivity scanner (TCS)

I.   Introduction: TSC Principle and Measurement Results 2

II.   Thermal conductivity profiles 4

III. Theory and Calculation Results 7

     III.1. Geometric Method 8

     III.2. Arithmetric Method 9

     III.3. Harmonic Method..9

IV. Matrix thermal conductivity (λ_{ma}) 10

V.   Error calculation 10

VI. Conclusion 11

VII. References..13

I. Introduction: TCS Principle and Measurement Results

The thermal conductivity scanner (or TCS) measures the thermal conductivity (W/m·K) via optical scanning method. In Figure 1 is a picture of the measurement shown. The thermal conductivity is a material property. High values are used for cooling systems to transport heat away from the material in a short time (e.g. fringes); low values are used as insulators, e.g. thermos flasks.

Fig. 1: Thermal conductivity scanner (TCS) on the E.ON Energy Research Center.

With this method the heater and the detectors where moved along the sample from the right to the left. A sensor measures within the moving first the "cold" conductivity of the sample. Than the heater follows and a last sensor measures after the heating of 25°C the conductivity of the sample again. Before and after the sample are two reference sources laid with a defined thermal conductivity. These reference sources are need for the sensor calibration, too. To avoid errors there has always to be a gap of a few centimeters between the reference blocks and the measured sample. For the measuring the samples and the reference sources have to be colored with a thick black line to avoid overheating and reflection by lighter samples. The opening has to be covered completely by the painted part of the sample. To control this was a mirror underneath the apparatus.

The used core samples are G1, a black stone with small mica particles and bigger white quartz inclusions, and G2, a greenish sandstone with small particles and lower compaction, from former measurements. As a third sample get measured a heterogeneous metamorphic rock, which is coarse-grained with a lighter, colorful or even brownish colour. The measured values for the samples are shown in table 1. Every sample is measured twice in each direction. G1 and G2 are totally dry and saturated measured. The metamorphic rock gets measured just dry, but three times in different planes of the halved core, marked with a black paint. The marked values for G1 need to be corrected, because on the core sample was a bit of the black colour gone. To avoid errors, these values have to be disregard.

	λ [W/m·K]	min	max	G [%]	Inhomogen.
G1 dry	3.087	2.899	3.256	2.620	0.116
G1 dry	3.079	2.923	3.297	2.358	0.122
Average	3.083	2.911	2.911	2.489	0.119
G2 dry	2.789	2.713	2.971	2.041	0.093
G2 dry	2.775	2.636	2.894	1.872	0.093
Average	2.782	Difference	2.675	1.957	0.093
G1 satura.	**2.915**	**2.762**	**3.056**	**2.416**	**0.101**
G1 satura.	3.017	2.865	3.140	2.529	0.091
G1 satura.	3.015	2.865	3.118	2.648	0.084
Average	2.782	Difference	2.865	1.957	0.088
G2 satura.	3.047	2.900	3.270	2.655	0.12
G2 satura.	3.108	2.977	3.240	1.692	0.084
Average	3.016	Difference	2.939	2.589	0.102
Metamor. 1	2.693	2.448	2.862	3.833	0.153
Metamor. 2	2.732	2.532	2.887	3.047	0.130
Metamor. 3	2.700	2.473	2.934	3.483	0.171
Average	2.708	Difference	2.484	3.454	0.151

Tab.1: Measured values for the rock sample from the TCS. Not the whole rock sample was taken for the average because of appeared boundary effects. The standard deviation is given as G [%].

II. Thermal conductivity and profiles

The changing in thermal conductivity is normally caused by change in mineral composition und therefore it is caused bei inhomogeneties. The metamorphic sample and sample G1 show inhomogeneties, which can be observed in figure 2,3 and 6 at some changes in the thermal conducitivity of about 0.2-0.3 W/m·K. Even sample G2 shows some inhomogeneties, because there is an increase in thermal conductivity over the position, seen in figure 4 and 5, but they are less extreme change in mineral composition than in the other samples.

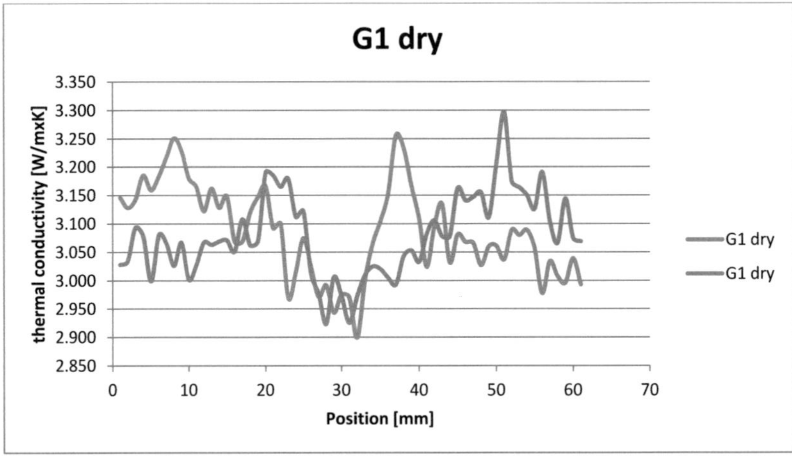

Fig. 2: Plotted profile for the dried sample G1. The measurement was done forward and backward by rotating the sample.

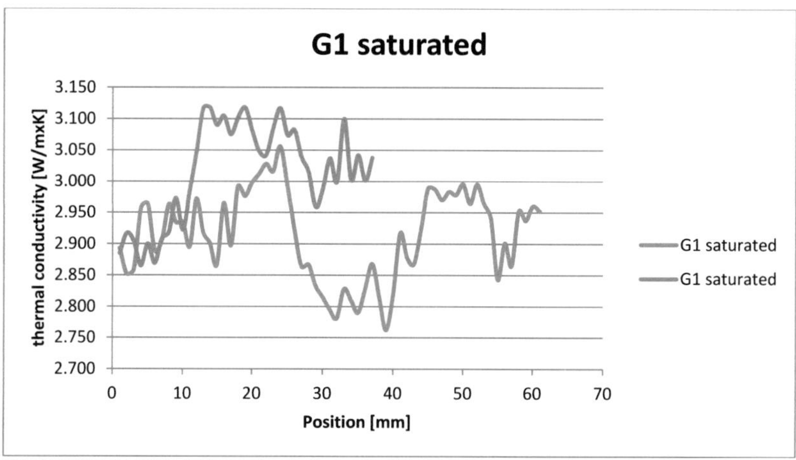

Fig. 3: Plotted profile for the saturated sample G1. The measurement was done forward and backward by rotating the sample.

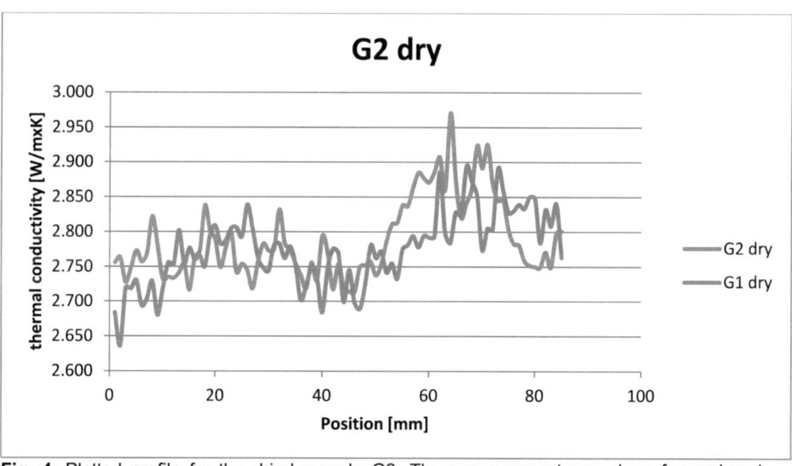

Fig. 4: Plotted profile for the dried sample G2. The measurement was done forward and backward by rotating the sample.

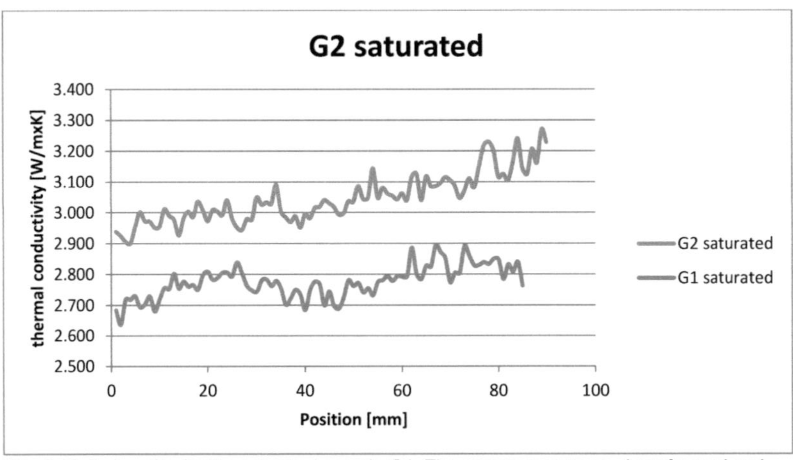

Fig. 5: Plotted profile for the saturated sample G1. The measurement was done forward and backward by rotating the sample.

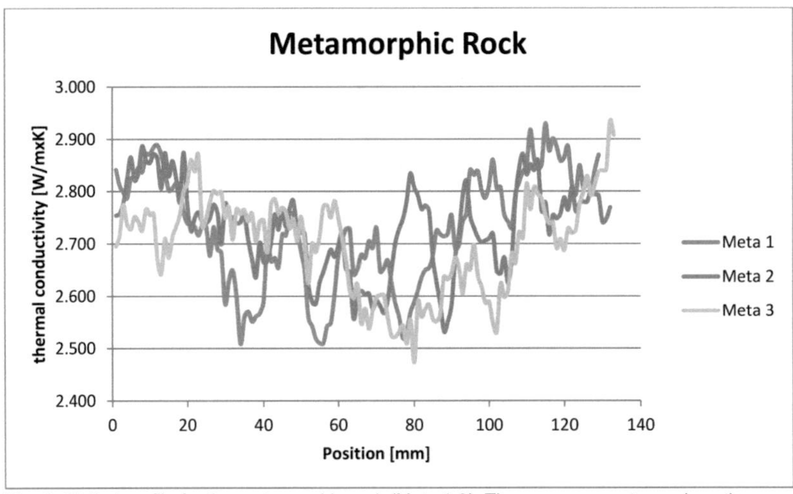

Fig. 6: Plotted profile for the metamorphic rock (Meta 1-3). The measurement was done three times in different planes of the roundish core, marked with a black paint.

III. Theory and Calculation Results

With the thermal conductivity values the porosity of the core samples can be calculated by three different approaches and the following equations: The arithmetic, the harmonic and the geometric mean approach.

$$\lambda_{arithmetic,sat} = \Phi\,\lambda_{water} + (1-\Phi)\,\lambda_{ma} \leftrightarrow \lambda_{sat} - \Phi\,\lambda_{water} = (1-\Phi)\,\lambda_{ma}$$

$$(1)$$

$$\lambda_{arithemtic,dry} = \Phi\,\lambda_{air} + (1-\Phi)\,\lambda_{ma} \leftrightarrow \lambda_{dry} - \Phi\,\lambda_{air} = (1-\Phi)\,\lambda_{ma}$$

$$(2)$$

$$\lambda_{harmonic,dry} = \frac{1}{\frac{1-\Phi}{\lambda_{ma}}+\frac{\Phi}{\lambda_{air}}} \leftrightarrow \frac{1}{\lambda_{sat}} = \frac{1-\Phi}{\lambda_{ma}} + \frac{\Phi}{\lambda_{air}} \leftrightarrow \frac{1}{\lambda_{sat}} - \frac{\Phi}{\lambda_{air}} = \frac{1-\Phi}{\lambda_{ma}}$$

$$(3)$$

$$\lambda_{harmon,sat} = \frac{1}{\frac{1-\Phi}{\lambda_{ma}}+\frac{\Phi}{\lambda_{water}}} \leftrightarrow \frac{1}{\lambda_{sat}} = \frac{1-\Phi}{\lambda_{ma}} + \frac{\Phi}{\lambda_{water}} \leftrightarrow \frac{1}{\lambda_{sat}} - \frac{\Phi}{\lambda_{water}} = \frac{1-\Phi}{\lambda_{ma}}$$

$$(4)$$

$$\lambda_{geometric,saturated} = \lambda_{water}^{\Phi} \cdot \lambda_{ma}^{1-\Phi} \leftrightarrow \frac{\lambda_{sat}}{\lambda_{water}^{\Phi}} = \lambda_{ma}^{1-\Phi}$$

$$(5)$$

$$\lambda_{geometric,dry} = \lambda_{air}^{\Phi} \cdot \lambda_{ma}^{1-\Phi} \leftrightarrow \frac{\lambda_{dry}}{\lambda_{air}^{\Phi}} = \lambda_{ma}^{1-\Phi}$$

$$(6)$$

Where λ_i is the thermal conductivity of the rock and not known, the individual approaches have to be solved by using two equations that air and water are present in the pore spaces. The values for this fluid thermal conductivities are at 25°C λ_{air} = 0.024 W/m·K and λ_{water} = 0.606 W/m·K (Lide, 2003). These values are used to calculate the dry and saturated sample. The porosity for G2 is 3.1 % and for G1 it is -1.3 % (= 0 %) for the geometric method (9), for the arithmetic method (7) G2 = 51 % and G1 = -21 % (=0%) and for the harmonic method (8) the porosity values are G2 = 0.0864 % and G1 = -0.034 % (=0%).

The porosity can also be determined with these three equations by conversion:

$$\Phi_{arithmeitc} = \frac{\lambda_{sat} - \lambda_{dry}}{\lambda_{water} - \lambda_{air}}$$

$$(7)$$

$$\Phi_{harmonic} = \frac{\frac{1}{\lambda_{sat}} - \frac{1}{\lambda_{dry}}}{\frac{1}{\lambda_{water}} - \frac{1}{\lambda_{air}}}$$

$$(8)$$

$$\Phi_{geometric} = \log\left(\frac{\lambda sat}{\lambda dry}\right) \cdot \frac{1}{(\log(\lambda_{water}) - \log(\lambda_{air}))}$$

(9)

III.1 Geometric Method – based on equation (5) and (6)

A calculation example for G2 dry is given below (using λ_{air}):

$$\Rightarrow \quad 2.782\,\frac{W}{m \cdot K} = \lambda^n \cdot 0.024\,\frac{W}{m \cdot K} \cdot \frac{1}{0.024\;W/m \cdot K^n}$$

$$\Rightarrow \quad 2.782\,\frac{W}{m} \cdot K = \left(\sqrt[n]{5.307\,\frac{W}{m \cdot K} \cdot 0.58\,\frac{W}{m \cdot K}}\right)^n \cdot 0.024\;W\,\frac{W}{m \cdot K} \cdot \frac{1}{0.024^n}$$

$$\Rightarrow \quad 2.782\,\frac{W}{m \cdot K} = 5.307\,\frac{W}{m \cdot K} \cdot 0.58\,W/m \cdot K^n \cdot 0.024\,\frac{W}{m \cdot K} \cdot \frac{1}{0.024^n}$$

$$\Rightarrow \quad \frac{2.782\,\frac{W}{m \cdot K}}{5.307\,\frac{W}{m \cdot K} \cdot 0.024\,\frac{W}{m \cdot K}} = \left(\frac{0.58\,\frac{W}{m \cdot K}}{0.24\,\frac{W}{m \cdot K}}\right)^n$$

$$\Rightarrow \quad 21.842\,\frac{W}{m \cdot K} = \left(\frac{0.58\,\frac{W}{m \cdot K}}{0.24\,\frac{W}{m \cdot K}}\right)^n$$

$$\Rightarrow \quad n = 0.97 \rightarrow 0.97 \cdot 100 = 97\%$$

$$\Rightarrow \quad 1 - n = 0.03 \rightarrow 0.03 \cdot 100 = 3\%$$

$$\Rightarrow \quad \sqrt[n]{115.917}\,\frac{W}{m \cdot K} \rightarrow \lambda = 3.24\,\frac{W}{m \cdot K}$$

Based on this example the calculation results are G2 saturated (using λ_{water}) = $3.24\,\frac{W}{m \cdot K}$, G1 dry (using λ_{air}) = $2.897\,\frac{W}{m \cdot K}$ and G1 saturated (using λ_{water}) = $2.901\,\frac{W}{m \cdot K}$.

III.2 Arithmetic Method – based on equation (1) and (2)

A calculation example for G2 dry is given below:

$$2.782\,\frac{W}{m\cdot K} = n\cdot\lambda + 0.024 - 0.024\,\frac{W}{m\cdot K}\cdot n$$

$$\Rightarrow\quad \frac{2.758\frac{W}{m\cdot K}+0.024\frac{W}{m\cdot K}}{n} = \lambda$$

$$\Rightarrow\quad 2.782\,\frac{W}{m\cdot K} = 2.498\,\frac{W}{m\cdot K} + 0.58\,\frac{W}{m\cdot K}\cdot n + 0.024\,\frac{W}{m\cdot K} - 0.024\,\frac{W}{m\cdot K}\cdot n$$

$$\Rightarrow\quad 2.782\,\frac{W}{m\cdot K} = 2.522\,\frac{W}{m\cdot K} + 0.556\,\frac{W}{m\cdot K}\cdot n$$

$$\Rightarrow\quad 0.26\,\frac{W}{m\cdot K} = 0.556\,\frac{W}{m\cdot K}\cdot n$$

$$\Rightarrow\quad n = 0.47 \rightarrow 0.47\cdot 100 = 47\%$$

$$\Rightarrow\quad 1-n = 0.53 \rightarrow 0.53\cdot 100 = 53\%$$

$$\Rightarrow\quad \lambda = \frac{2.758\frac{W}{m\cdot K}+0.024\frac{W}{m\cdot K}\cdot n}{n} = \boldsymbol{5.653\,\frac{W}{m\cdot K}}$$

Based on this example the calculation results are G2 saturated (using λ_{water}) = $5.651\frac{W}{m\cdot K}$, G1 dry (using λ_{air}) = $6.267\,\frac{W}{m\cdot K}$ and G1 saturated (using λ_{water}) = $5.44\frac{W}{m\cdot K}$.

III.3 Harmonic Method – based on equation (3) and (4)

The results for the samples using the harmonic method are $G2_{saturated}$ = $3.0889\frac{W}{m\cdot K}$, $G2_{dry}$ = $3.0889\frac{W}{m\cdot K}$, $G1_{saturated}$ = $2.957\,\frac{W}{m\cdot K}$ and $G1_{dry}$ = $2.147\frac{W}{m\cdot K}$. The used equation is shown above.

IV. Matrix thermal conductivity (λ_{ma})

With the porosity calculation of the Archimedes' method and the geometrical model the matrix thermal conductivity (λ_{ma}) can be calculated with the following equation:

$$\lambda_{geo} = \lambda_{ma}^{1-\phi} \; x \; \lambda_{fl}^{\phi} \quad \leftrightarrow \quad \frac{\lambda_{geo}}{\lambda_{fl}^{\phi}} = \lambda_{ma}^{1-\phi} \quad \leftrightarrow \quad \sqrt[1-\phi]{\frac{\lambda_{geo}}{\lambda_{fl}^{\phi}}} = \lambda_{ma} \tag{10}$$

Where λ_{TC} are the measured conductivity values seen above in table1, $\lambda_{porosity}$ are the literature values again for $\lambda_{air} = 0.024 \, \frac{W}{m \cdot K}$ and $\lambda_{water} = 0.606 \frac{W}{m \cdot K}$. The porosity values for sample G1 = 0.08% and G2 = 12% are given.

The results for this calculation are G2$_{dry}$ = 5.319$\frac{W}{m \cdot K}$, G2$_{saturated}$ = 3.841$\frac{W}{m \cdot K}$,

G1$_{dry}$ = 3.084$\frac{W}{m \cdot K}$ and G1$_{saturated}$ = 2.961$\frac{W}{m \cdot K}$.

V. Error calculation

The gaussian error propagation has been used to calculate the total $\Delta\phi$ error. The used equations for the arithmetic, harmonic and geometric values are shown in the following:

$$\Delta\phi_{ari} \left(\lambda_{sat}, \lambda_{dry} \right) = \sqrt{\left(\left(\frac{1 - \lambda_{dry}}{\lambda_{H2O} - \lambda_{air}} \right) x \, \Delta\lambda_{sat} \right)^2 + \left(\left(\frac{\lambda_{sat} - 1}{\lambda_{H2O} - \lambda_{air}} \right) x \, \Delta\lambda_{dry} \right)^2}$$

$$\tag{11}$$

$$\Delta\phi_{har} \left(\lambda_{sat}, \lambda_{dry} \right) = \sqrt{\left(\left(\frac{-\lambda_{sat}^{-2} - \lambda_{dry}^{-1}}{\lambda_{h2o}^{-1} - \lambda_{air}^{-1}} \right) x \, \Delta\lambda_{sat} \right)^2 + \left(\left(\frac{\lambda_{sat}^{-1} + \lambda_{dry}^{-2}}{\lambda_{h2o}^{-1} - \lambda_{air}^{-1}} \right) x \, \Delta\lambda_{dry} \right)^2}$$

$$\tag{12}$$

$$\Delta\phi_{geo} \left(\lambda_{sat}, \lambda_{dry} \right) = \sqrt{ \begin{aligned} &\left(\left(\left(\frac{1}{\lambda_{sat} * \ln(10)} \right) - \log(\lambda_{dry}) \right) * \frac{1}{\log(\lambda_{H2O}) - \log(\lambda_{air})} \right) * \Delta\lambda_{sat})^2 + \\ &\left(\left(\left(\lambda_{sat} - \frac{1}{\lambda_{dry} * \ln(10)} \right) * \frac{1}{\log(\lambda_{H2O}) - \log(\lambda_{air})} \right) * \Delta\lambda_{dry})^2 \end{aligned} }$$

$$\tag{13}$$

$\Delta\lambda_{sat}$ and $\Delta\lambda_{dry}$ are given by the mean standard deviation of the measurements shown in

table 1 under G [%].

The error of λ_{water} and λ_{air} is unknown. Therefore they have been used as true values.

The results for the error propagation calculation are shown in table 2.

	$\Delta\phi_{ari}$ [%]	$\Delta\phi_{geo}$ [%]	$\Delta\phi_{har}$ [%]
G1	1.524	0.92	0.037
G2	0.931	0.66	0.033

Tab. 2: Calculated results for the error calculation for the TCS experiment.

What can be observed in table 2 is that the highest error is caused by the arithmetic method, but even here it is just by about 1-1.5%. The lowest errors were got with the harmonic method where the errors are nearly 0%. Even the error propagation for the geometric method is acceptable.

VI. Conclusion

The thermal conductivity values for sample $G1_{dry}$ are higher than for $G2_{dry}$ by about 0.3 W/m·K, but for the saturated samples G2 is higher than G1. This caused due to the fact, that sample G2 has a higher porosity, than G1, because the pore water increases the thermal conductivity. This supports our assumption that G2 is a porous sandstone. Literature values[1] are between 2.1-2.3 W/m·K. In contrast G1 has no pores inside, so the thermal conductivity doesn't change with saturation. This leads to the assumption that sample G1 is a densly packed gabbro. In comparison to literature values[2] gabbro is up to 4 W/m·K, the measured values are slighlty below, but would fit. The high thermal conductivity values could be explained by conductive minerals inside the matrix, for e.g. iron-rich minerals like magnetite or fayalite. The metamorphic rock, which was measured at least, has slightly lower values for the thermal conductivity, than sample G2 and fit in comparison with literature values[3] of 1.9-4 W/m·K to a gneiss. This can be caused by iron-less minerals inside the matrix. But what has to take in account is, that the error is higher for metamorphic rock (average 3.454%) than for sample G2 (average $G2_{dry}$ 1.957 %, $G2_{sat}$ 2.589%). The error value for $G1_{dry}$ is 2.489% and for $G1_{sat}$ 1.957%. For the three measured samples the metamorphic rock has the highest error.

A comparison of the porosity values determined by the three different parameters with the given values determined by Archimedes' measurement is shown in table 3.

Porosity [%]				
Model	**arithmetic**	**harmonic**	**geometric**	**Table 1**
G1	-21 (=0%)	-0.034 (=0%)	-1.3 % (= 0 %)	0.08
G2	51	0.0864	3.1	12

Tab. 3: Comparison of the calculated porosity values for each sample and each method with the given values measured by Archimedes' method.

The table shows that the geometric method is most comparable with the given Archimedes' porosity. The arithmetic and harmonic values strongly deviate from the geometric porosity values. But even the geometric method does not fit completely to the Archimedes' porosity. This can be caused by measurement or calculation errors.

VII. References

- http://de.wikipedia.org/wiki/W%C3%A4rmeleitf%C3%A4higkeit (30.01.13) [1]

- www.materialarchiv.ch (29.01.13) [2]

- http://www.heizungsfinder.de/waermepumpe/sole-wasser/erdreich (30.01.13) [3]